Process Development Symposium 2024

Rosemont, Illinois, USA
4 – 6 June 2024

ISBN: 979-8-3313-0082-1

Printed from e-media with permission by:

Curran Associates, Inc.
57 Morehouse Lane
Red Hook, NY 12571

Some format issues inherent in the e-media version may also appear in this print version.

AIChE
120 Wall Street, FL 23
New York, NY 10005-4020

Phone: (800) 242-4363
Fax: (203) 775-5177

www.aiche.org

Additional copies of this publication are available from:

Curran Associates, Inc.
57 Morehouse Lane
Red Hook, NY 12571 USA
Phone: 845-758-0400
Fax: 845-758-2633
Email: curran@proceedings.com
Web: www.proceedings.com

TABLE OF CONTENTS

5-BIO-PROCESS DEVELOPMENT, PHARMACEUTICALS, ALTERNATIVE FUELS

Advancements Toward the Manufacturing Route for the Fungicide Haviza

CONFERENCE

Process Development Symposium

YEAR

2024

PROCEEDING

2024 Process Development Symposium

GROUP

2024 Process Development Symposium Program

SESSION

Designing Sustainable Processes and Products

TIME

Wednesday, June 5, 2024 - 10:15am to 10:45am

Authors

Davis, T. - Presenter

Despite significant recent advances in agricultural solutions, diseases caused by fungal pathogens lead to estimated crop losses of 10-20% each year. Asian soybean rust, caused by the fungi *Phakopsora pachyrhizi* (PHAKPA), is a disease which leads to widespread loss of soybean crops. The fungicide Haviza (metarylpicoxamid) has been developed by Corteva Agriscience to control this pathogen. Metarylpicoxamid is a stereochemically complex active ingredient. One of the isomers was found to be most active and was chosen for development. This resulted in the need for a highly selective and efficient synthetic process to deliver a product capable of maximum disease control and minimal application rates. Additionally, sustainability and cost were important drivers for process development. Details about the selection and development of the synthetic routes for Haviza will be discussed.

Effects of Soybean Varieties on Life-Cycle Greenhouse Gas Emissions of Biodiesel and Renewable Diesel

CONFERENCE

Process Development Symposium

YEAR

2024

PROCEEDING

2024 Process Development Symposium

GROUP

2024 Process Development Symposium Program

SESSION

Designing Sustainable Processes and Products

TIME

Wednesday, June 5, 2024 - 11:15am to 11:45am

Authors

Li, Y. - Presenter, Corteva Agriscience

The transportation sector accounted for about 29% of the total US greenhouse gas (GHG) emissions owing to its high reliance on fossil fuels. Adoption of biofuels is one of the viable options to decarbonize the hard-to-electrify sectors. Biodiesel (BD) and renewable diesel (RD) are two major biomass-derived diesel fuels which are capable for blending in or alternating fusil diesel in long-haul road trucks and maritime ships.
This study presents a life-cycle analysis (LCA) of the greenhouse gas (GHG) emissions of BD and RD produced from soybean varieties with altered oil profiles and meal composition. The varieties evaluated include those with high

oleic oil content (78% of oil being oleic oil vs. the current content of 25%), high lipid content (25% vs. the current content of 19%) or high protein content (45% vs. the current content of 35%). The results suggest that high-oleic soybean oil could reduce GHG emissions of RD by 1 g CO2e/MJ by reducing hydrogen consumption in the hydro-treating stage of RD production. With the default allocation method (mass-based allocation between oil and meal), changes in lipid and protein contents have negligible impacts on GHG emissions of BD/RD pathways. Since high-protein soybean meal has a higher nutrient value, we employed the market-based allocation for sensitivity analysis. Assuming a market price of $1.17/dry kg and an average of two substitute feeds with similar protein contents (corn gluten and fish meal), the allocation factor for soybean meal increases from 63 to 81%. Compared with the results for the case with commodity soybean using mass allocation, GHG emissions could be 1.9 and 1.8 g CO2e/MJ lower for BD and RD, respectively. In addition to product-level LCA, we employed a land-based LCA to evaluate the effect of soybean properties on product slates with the same land area and the associated displacement of conventional products. The results suggest that high-lipid soybeans may offer more GHG savings than other varieties as the higher BD/RD yield replaces more fossil diesel.

Modeling Solvent-Based Dissolution-Precipitation Recycling of PET with Process Simulation

CONFERENCE

Process Development Symposium

YEAR

2024

PROCEEDING

2024 Process Development Symposium

GROUP

2024 Process Development Symposium Program

SESSION

Designing Sustainable Processes and Products

TIME

Wednesday, June 5, 2024 - 10:45am to 11:15am

Authors

Peralta, A. - Presenter, Chemstations Inc

Solvent based dissolution-precipitation is an emerging chemical recycling technology that has the potential to close the loop for recycling Polyethylene terephthalate (PET), a common polymer resin used in consumer packaging. The solvent Gamma-Valerolactone (GVL) was selected for this process because it can be obtained from renewable biomass. Three different methods for producing high-quality PET from post-consumer PET waste were modeled using process simulation software: (1) adding an anti-solvent, (2) solvent evaporation, and (3) cooling the solution to precipitate dissolved polymer.

Process models were simulated with and without heat integration to determine the impact to process efficiency and cost savings.

Enablement of Photochemistry Processes Through Technology and Cross Functional Teams

CONFERENCE

Process Development Symposium

YEAR

2024

PROCEEDING

2024 Process Development Symposium

GROUP

2024 Process Development Symposium Program

SESSION

Innovation & New Technologies

TIME

Wednesday, June 5, 2024 - 1:30pm to 2:00pm

Authors

Nagy, K. - Presenter, Snapdragon Chemistry

The demands for bringing commercially viable processes to the market have become more difficult in recent years, with ever increasing emphasis on sustainability, cost, and time to market. In the pharmaceutical space, molecules have become more complex, timelines are continuously compressed, and purity requirements continue to tighten. Overcoming these challenges requires both the development and application of new chemical processing technologies and the efforts of not only chemists and chemical engineers, but also mechanical, automation, and software engineers. In this talk, Iâ€™ll review some of the ways in which Snapdragon has approached these grand challenges by leveraging automation and advanced manufacturing technologies to facilitate early phase GMP deliveries of complex molecules with a case study focus on photochemistry process and photoreactor design. While photochemistry offers chemists unique ways to make molecules without requiring high thermal inputs, deployment of commercial photoreactor processes has remained rare due to their complexity to build and operate. Design of a photochemistry process requires a detailed understanding of photophysics and reaction kinetics, and photoreactor design requires input from a broad array of scientists and engineers to ensure that hazards associated with high voltage systems and high intensity light are mitigated while tightly controlling light irradiance, temperature of the process fluid, and temperature of the light source. Iâ€™ll review the history of development of photochemistry reactor technology at Snapdragon, how we have approached and overcome these operational challenges to build a 5 kW photochemical reactor platform, and provide case studies where the technology has been successfully deployed and proven transformational.

Solvent Extraction for Uranium Recovery and Impurity Rejection

CONFERENCE

Process Development Symposium

YEAR

2024

PROCEEDING

2024 Process Development Symposium

GROUP

2024 Process Development Symposium Program

SESSION

Innovation & New Technologies

TIME

Wednesday, June 5, 2024 - 2:30pm to 3:00pm

Authors

Stammitti, A. - Presenter, Insight RND
Faucher, S., McMaster University

A novel uranium beneficiation process has been developed by Insight R&D inc. for Global Atomic Corporation (GAC) to process ores produced by the DASA uranium mine in Niger. The process consists of comminution, acid pugging and leaching, solvent extraction, and precipitation. Solvent Extraction (SX) is used to selectively separate uranium from impurities in the leach solution. A tertiary amine extracts the uranyl cation into the organic phase but also has a tendency to carry over some impurities. Scrubbing is utilized to address some of this carry-over followed by stripping. Stripping transfers the extracted cations from the loaded organic to an aqueous phase. In this work two stripping agents (sodium chloride and sodium carbonate) were explored at pilot-scale and their advantages and disadvantages assessed for downstream product quality. Upon stripping agent selection, the SX reagent concentrations were optimized utilizing METSIM® to minimize carry over of acid and impurities from the front to the back end of the SX circuit.

Sustainable Aviation Fuels

CONFERENCE

Process Development Symposium

YEAR

2024

PROCEEDING

2024 Process Development Symposium

GROUP

2024 Process Development Symposium Program

SESSION

Innovation & New Technologies

TIME

Wednesday, June 5, 2024 - 2:00pm to 2:30pm

Authors

Serban, M. - Presenter, Honeywell UOP

Sustainable Aviation Fuels (SAF) demand is expected to accelerate in the coming years, as government targets and mandates will drive its further adoption.
Honeywellâ€™s UOP Ecofining™ technology, commercialized in 2013, has enabled significant capacity of renewable diesel and SAF product, however, achieving targeted SAF production from only Hydroprocessed Esters and Fatty Acids (HEFA) feedstocks is challenging due to the continued availability of these feeds. Other production pathways enabling the use of alternate renewable feedstocks are needed to satisfy the growing SAF demand.

This presentation will cover the latest additions to Honeywellâ€™s UOP SAF portfolio, i.e., UOP Ethanol-to-Jet (ETJ) Processâ„¢ and UOP eFining™ technologies, commercialized in 2022 and 2023, respectively.

Ethanol, an abundant and low-cost bio-feedstock, and methanol, produced from carbon dioxide and green hydrogen, can be converted to Honeywell Green JetTM fuel through the ETJ and eFining processes with significantly lower greenhouse gas (GHG) emissions on a total lifecycle basis, compared to petroleum-based jet fuel. Our innovative ETJ and eFining processes leverage more than a decade of experience and UOPâ€™s proven innovation in the space, offering high jet yield output, low CAPEX & OPEX, reduced GHG emissions and the ability to commercialize at scale. Additionally, these low-technology-risk processes are flexible for converting renewable and waste-derived alcohols into Honeywell Green DieselTM fuel.

Keywords: Renewable Fuels, Sustainable Aviation Fuels, Catalysts Development, Process Development

Unlocking Efficiency and Innovation Through Intensified Process Design for Specialty Chemicals

CONFERENCE

Process Development Symposium

YEAR

2024

PROCEEDING

2024 Process Development Symposium

GROUP

2024 Process Development Symposium Program

SESSION

Innovation & New Technologies

TIME

Wednesday, June 5, 2024 - 3:00pm to 3:30pm

Authors

Al Azri, N. - Presenter

The Process Innovation Team at Lubrizol plays a critical role in leading process intensification research, exploring novel manufacturing technologies, and developing sustainable solutions for specialty chemicals processing. The team was formed followinga successful collaboration between Lubrizol, the University of Pittsburgh, the RAPID Manufacturing Institute, and the Department of Energy in 2017. This collaboration allowed for the implementation of advanced modular manufacturing technologies for the continuous processing of dispersants at Lubrizol. As a result, the team became an integral part of the RAPID Institute and gained the ability to drive global

trends in process development research and batch-to-continuous transition technology.
In this presentation, we will provide valuable insights into the Process Innovation Team's pivotal role at Lubrizol. We will discuss the team's achievements and its commitment to driving innovation, sustainability, and efficiency in the specialty chemicals industry. We will also take this opportunity to review lessons learned from the transition from batch to continuous processes and Modular Chemical Process Intensification (MCPI). The audience will gain a deeper understanding of how Lubrizol is driving advances to manufacturing technology through the team's five key focus areas: process intensification, modular manufacturing, sustainability, process modeling, and external partnerships to accelerate speed to market and de-risk scale up technologies.

Capitalizing on Pilot Scale for a Mature Process Technology Scale-up

CONFERENCE

Process Development Symposium

YEAR

2024

PROCEEDING

2024 Process Development Symposium

GROUP

2024 Process Development Symposium Program

SESSION

Batch vs. Continuous, Scale-Up & Scale-Down

TIME

Thursday, June 6, 2024 - 9:30am to 10:00am

Authors

Schultz, L. - Presenter
Patil, R., Sabic Innovative Plastics

SABIC has been a leading solutions provider of differentiated and novel offerings in the world of amorphous polymers for more than two decades. As new LNPâ„¢ copolymers are designed or invented, each comes with its own unique set of challenges in scale-up translation to manufacturing. Complexity in a new resin process design creates additional challenges. Each challenge, however, presents an opportunity to examine fundamentals-based fit between a new copolymer and a unit operation. SABICâ€™s state-of-the-art pilot facility has enabled in-depth studies of relationship among process, product, and quality CTQâ€™s. The end result is a very mature scale-up package that saves production line time and minimizes losses.

This presentation will focus on one such challenging scale-up that was defined by a mismatch of the copolymer resin process between the plant sections of reaction and isolation. The polymerization reaction required unique pH conditions for a part of the reaction. This limited the polymerization step by restricting the addition rate of other raw materials. The isolation section must take the output of the reaction section and requires fast monomer incorporation under unique pH conditions to move forward in stable operation. Pilot scale explorations deliberately investigated the key reaction parameters at the extreme conditions. Resulting materials underwent comprehensive product performance testing and various analytical characterizations. The outcomes were applied to devise optimal reaction operation regime and a detailed manufacturing scale-up package. When the very first scale-up production runs were conducted, the plant achieved safe and stable operation for the entire duration and produced impressively high yields of the finished product.

Process Development & Scale-Up of Novel Organosilicon Coupling Agent

CONFERENCE

Process Development Symposium

YEAR

2024

PROCEEDING

2024 Process Development Symposium

GROUP

2024 Process Development Symposium Program

SESSION

Batch vs. Continuous, Scale-Up & Scale-Down

TIME

Thursday, June 6, 2024 - 9:00am to 9:30am

Authors

DePierro, M. - Presenter

A phase transfer catalyzed (PTC) reaction process was recently developed for a novel organosilane coupling agent. Using a geometrically similar lab reactor and similar mixing to the plant we developed and successfully scaled up the process. Understanding the rate limiting factors for this system was critical to optimizing reaction conditions. Mixing intensity, catalyst structure, and solvent selection were important considerations for scale up and overall efficiency of the process. The presence of trace polar protic species strongly influenced the reaction behavior, crystallization, and filtration performance.

An R&D Perspective on the Scale-Up of Heterogeneous Catalysts

CONFERENCE

Process Development Symposium

YEAR

2024

PROCEEDING

2024 Process Development Symposium

GROUP

2024 Process Development Symposium Program

SESSION

Batch vs. Continuous, Scale-Up & Scale-Down

TIME

Thursday, June 6, 2024 - 10:00am to 10:30am

Authors

Sussman, V. - Presenter

The scale up of heterogeneous catalysts is an important step in enabling the profitable industrial production of chemicals to benefit society. The first part of this talk will discuss some of the common pitfalls and challenges associated with translating laboratory discoveries into practical commercial solutions, highlighting both technical and business challenges that researchers may encounter as they seek to reduce their inventions to practice. In the second part of the talk, we will discuss how these types of challenges were encountered and addressed during the development and scale up of a hydrogenation catalyst at Dow and describe how careful application of catalysis and reaction engineering fundamentals was critical to enabling the successful production of the catalyst at scale.

Integrating Functional Principal Component Analysis with Data-Rich Experimentation for Enhanced Drug Substance Development

CONFERENCE

Process Development Symposium

YEAR

2024

PROCEEDING

2024 Process Development Symposium

GROUP

2024 Process Development Symposium Program

SESSION

Modeling, Computation, and Data Analytics in Process Development

TIME

Thursday, June 6, 2024 - 10:50am to 11:20am

Authors

McMullen, J. - Presenter, Merck & Co.

The combination of laboratory automation and data-rich experimentation has led to significant improvements in experimental efficiency, reproducibility, and throughput compared to conventional methods. These transformative technologies have proven particularly valuable in drug substance process characterization studies, where the ability to rapidly generate volumes of reaction data to maximize process understanding for ever-increasing regulatory expectations is of paramount importance. These highly dense data sets offer significant potential when coupled with data-driven modeling to quantify reaction dynamics and sensitivities for process knowledge and optimization. To ensure that this analysis is applied throughout drug substance reaction development, identifying a data-driven modeling approach capable of describing diverse reaction trends is of critical need. In this study, functional principal component analysis for data-driven reaction modeling was performed to highlight its applicability for drug substance development. To demonstrate this methodology, we employed automated reactor and sampling technologies in the process characterization studies of a heterogeneous fluorination reaction using gaseous trimethylamine. These data-rich experiments were structured according to a 24 full factorial design of experiment, each comprising at least 12 reaction time samples. By applying functional principal component analysis to various reaction responses, the optimal design space for manufacturing operations was easily identified.

Understanding a Reactive Crystallization Through Thermodynamic and Kinetic Modeling

CONFERENCE

Process Development Symposium

YEAR

2024

PROCEEDING

2024 Process Development Symposium

GROUP

2024 Process Development Symposium Program

SESSION

Modeling, Computation, and Data Analytics in Process Development

TIME

Thursday, June 6, 2024 - 11:20am to 11:50am

Authors

Bhamidi, V. - Presenter

Process development of active ingredients and intermediates relevant to the agrochemical and pharmaceutical industry often involves reactive crystallization. These synthesis schemes are usually complicated by the relative influences of chemical reaction and crystallization on the outcome of the process. Often, the solution conditions that result in a good reaction with desirable product formation and high yield are not the optimal conditions for isolating the formed product in a solid form with high recovery. Here, we discuss a case study on the synthesis of an agrochemical intermediate wherein the target molecule is prone to oiling out as soon as the reaction progresses, thus complicating the subsequent execution of the process scheme.

To address this situation, systematic experiments that probed the reaction kinetics and crystallization thermodynamics (without mutual influence) were carried out. These data were used to build a reaction kinetic model and a phase diagram model for the compound. Combining these two modules into a master reactive crystallizer model, the proximity and potential intersection of the process path and the liquid-liquid phase separation (oiling out) boundaries were readily characterized. The analysis revealed that in the existing process scheme the reaction and crystallization were separated in time resulting in a rapid crystallization event during the later part of the process. Alternate process schemes that can reduce / eliminate oiling out and distribute crystallization evenly throughout the time course of the reaction were identified. The approach discussed here highlights the effectiveness of first-principles modeling in providing the team with critical insights needed for identifying a robust process scheme, thus reducing empiricism in process development.

Development of pDADMAC Flocculation Harvest for Intensified Fed-Batch Process

CONFERENCE

Process Development Symposium

YEAR

2024

PROCEEDING

2024 Process Development Symposium

GROUP

2024 Process Development Symposium Program

SESSION

Bio-Process Development, Pharmaceuticals, Alternative Fuels

TIME

Thursday, June 6, 2024 - 2:00pm to 2:30pm

Authors

Li, Z. - Presenter, University of Central Florida

The separation of cells, cell debris, and host cell related impurities from the target product in biologics manufacturing becomes increasingly challenging as cell density and titer in the upstream bioreactor rise. Traditional filtration methods require large membrane areas or multiple stages, leading to facility fit issues and increased costs. To address these challenges, flocculation with cationic polymers has been explored, but certain key questions regarding its successful application and scale-up remain unanswered.
This study aims to investigate the use of pDADMAC flocculation for consistent flocculate morphology from bench to pilot scale and effective removal of

pDADMAC during downstream processing. Cell broth obtained from an intensified upstream process, capable of achieving a peak cell density of 38 million/mL, was used. Real-time imaging with Easyviewer was employed to quickly screen flocculation conditions based on cell density, viability, and pDADMAC concentrations. EasyViewer, both in the bioreactor and flow-cell setup, proved to be a versatile tool for evaluating flocculate particle morphology across different scales.

Successful application of pDADMAC flocculation at a 500L pilot scale achieved a loading of over 130 L/m2 with minimal pressure buildup using only one stage of depth filtration. Additionally, this study provides the first quantitative data, obtained through reverse phase HPLC, for pDADMAC mass balance analysis. The majority of residual pDADMAC after flocculation was found to be removed during the high salt wash in Protein A chromatography. Extensive characterization of flocculated and non-flocculated samples demonstrated that the inclusion of the flocculation step reduces impurity levels at the early stage of downstream purification. Compared to the control group, a 40% reduction in HCP concentrations (measured by ELISA), a 27% reduction in the number of HCP species (measured by mass spectrometry), and complete removal of DNA were observed. The study also addressed potential risks of disulfide bond reduction and particle formation after harvest due to residual pDADMAC.

These results expand the downstream purification toolbox for harvesting cell broth with high cell density and low viabilities. Furthermore, they provide valuable insights into the integration of flocculation-depth filtration with the current platform process for biologics purification.

Process Development and Characterization for Impurity Control of a Dynamic Kinetic Resolution Reaction Using Data Rich Experimentation

CONFERENCE

Process Development Symposium

YEAR

2024

PROCEEDING

2024 Process Development Symposium

GROUP

2024 Process Development Symposium Program

SESSION

Bio-Process Development, Pharmaceuticals, Alternative Fuels

TIME

Thursday, June 6, 2024 - 2:30pm to 3:00pm

Authors

Dubina, S. - Presenter
McCann, S., Merck and Co., Inc.
Vi, T., Merck & Co., Inc.
Schultz, V., Merck & Co., Inc.
Padivitage, N., Merck and Co., Inc.
DiRocco, D., Merck & Co., Inc.
McMullen, J., Merck & Co.

Telescoped processes with multiple reactions can streamline manufacturing in the commercial syntheses of APIs and their intermediates. However, robustness challenges may arise, as there tends to be no normalization like that of isolated intermediates. To mitigate risks caused by these challenges, Data Rich Experimentation (DRE) is utilized to characterize telescoped processes to better enhance understanding around process factor sensitivity. This presentation covers the development of a telescoped process involving Fluorination and Dynamic Kinetic Resolution (DKR) reactions in the commercial synthesis of an API intermediate before scaling up and tech transferring to larger scales.

The current filed process for Belzutifan involves a chemocatalytic reduction with Ru. To be more sustainable and avoid the price volatility of precious metals, a post approval route that employed KRED was developed, streamlining the process to utilize a greener, safer, and more robust chemo-enzymatic through-process over the previous Ru-catalyzed route. Employing biocatalysis introduced new process challenges, including new reaction impurities and sensitivities to incoming materials from upstream reactions. Prior to scaling up to a commercial site, this DKR reaction was meticulously studied to assess the operable design space around applicable process variables. After designing a baseline procedure that maintained enzyme stability and provided the necessary reaction conversion, the operating ranges of the reaction parameters were refined through risk assessment for quality. Range finding experiments were performed using DRE systems to quickly and efficiently identify the boundaries of the design space. Results from these DRE screens reduced the number of parameters needed in subsequent experimentation. Process characterizations were performed according to a Design of Experiment (DoE) structure to statistically quantify the relationship between operating factors and reaction responses. These experiments were also performed with elements of a DRE strategy, using process analytical tools (PAT) and automated reactor sampling to generate reaction profiles to understand kinetics as well as end of reaction conditions. Data analysis generated practical understandings regarding pH control and KRED loading on the production of several reaction impurities. Laboratory data was bridged to the production environment through successful scale-up to a pilot plant facility. Key takeaways from the risk assessment work instilled confidence in scaling up to the manufacturing production plant.

AUTHOR INDEX

AIChE
120 Wall Street, FL 23
New York, NY 10005-4020

ISBN 979-8-3313-0082-1

CURRAN ASSOCIATES INC.
proceedings.com